It's A Rainforest

Larry Keen
RJ Pollack
James Copacino

Illustrations by Colleen Taylor

Copyright © 2022 by Pollack, Copacino and Keen
All Rights Reserved
ISBN: 978-8-218-10341-5

"It's A Rainforest" is the second children's book by the authors/lyricists, Larry Keen, RJ Pollack and James Copacino. Their first book "The Cat, The Fiddle and Me" was launched in 2019 in the form of a hardcover and an e-book.

Our wish is that you will once again enjoy the words, the music and the dazzling and fanciful illustrations by Colleen Taylor, which bring the story alive and capture the vibrant color of the Amazon Rainforest and the animals that live there. Singer, songwriter and entertainer, Larry Keen has spent a lifetime playing his special blend of folk, rock and country music from New York to LA and everywhere in between. The musical and lyrical writing team of RJ Pollack, James Copacino and Larry Keen, also known as RJ, Cino and Keener, are delighted to bring you another magical journey into the Amazon Rainforest.

"A half world away, where the Jaguar's play, every night as the world spins round"

It's a rainforest, a magic world for you and me.
But there will be no air to breathe if we cut down the trees,
listen mister can't you see.

It's a rainforest, where the animals and plants abound

The eagle's glide through the azure sky from dawn
'till the sun goes down

I turned on the TV and much to my surprise;
they were cutting down the jungle right before my eyes.

We don't need another parking lot or towers that touch the clouds. Give me the forest so green and air that is clean. If you listen, you can hear the sound.

It's a rainforest, a paradise for all to see,
but this beautiful world's in trouble, it's up to you and me.

It's a rainforest, way down in the Amazon. A half world away, where the jaguar's play, every night as the world spins round

High up in the canopy, as far as the eye can see

Where the Three Toed Sloth and the Scarlet McCaw,
live in harmony

Maybe I'm just a dreamer, but if we all believe, we can all join hands in Mother Nature's band in the shade of the Kapok Tree

It's a rainforest, a magic world for you and me

But there will be no air to breathe if we cut down the trees.
Listen mister can't you see

It's a rainforest, where the animals and plants abound

The eagles glide through the azure sky from dawn until the sun goes down

Rainforest Knowledge:

Rainforest's are Earth's oldest living ecosystems, with some surviving in their present form for at least 70 million years. They are incredibly diverse and complex, home to more than half of the world's plant and animal species—even though they cover just 6% of Earth's surface. This makes rainforest's astoundingly dense with flora and fauna. A 4-square-mile patch can contain as many as 1,500 flowering plants, 750 species of trees, 400 species of birds and 150 species of butterflies.

Rainforest's rich biodiversity is incredibly important to our well being and the well being of our planet. Rainforest's help regulate our climate and provide us with everyday products. Unsustainable industrial and agricultural development, however, has severely degraded the health of the world's rainforest's. Citizens, governments, intergovernmental organizations, and conservation groups are working together to protect these invaluable but fragile ecosystems.

Rainforest Structure:

Most rainforest's are structured in four layers. The top layer is the emergent layer. Beneath the emergent layer is the canopy, a deep layer of vegetation roughly 20 feet thick. The canopy blocks wind, rainfall and sunlight, creating a humid and dark environment below.

The understory layer is located several meters below the canopy. The understory is an even darker, stiller, and more humid environment. Animals call the understory home for a variety of reasons. Many take advantage of the dimly lit environment for camouflage.

The lowest layer is the forest floor layer, which is the darkest of all rainforest layers, making it extremely difficult for plants to grow. Leaves that fall to the forest floor decay quickly.

The Benefits of Rainforests:

Rainforest's are critically important to the well being of our planet. Tropical rainforest's encompass approximately 3 billion acres of vegetation and are sometimes described as the earth's thermostat.

Rainforest's produce about 20% of our oxygen and store a huge amount of carbon dioxide, drastically reducing the impact of greenhouse gas emissions. Massive amounts of solar radiation are absorbed, helping regulate temperatures around the globe. Taken together, these processes help to stabilize Earth's climate.

Rainforest's also help maintain the world's water cycle. More than 50% of precipitation striking a rainforest is returned to the atmosphere by evapotranspiration, helping regulate healthy rainfall around the planet. Rainforest's also store a considerable percentage of the world's freshwater, with the Amazon Basin alone storing one-fifth.

Rainforest's provide us with many products that we use every day. Tropical woods such as teak, balsa, rosewood, and mahogany are used in flooring, doors, windows, boatbuilding, and cabinetry. Fibers such as raffia, bamboo, kapok, and rattan are used to make furniture, baskets, insulation, and cord. Cinnamon, vanilla, nutmeg, and ginger are just a few spices of the rainforest. The ecosystem supports fruits including bananas, papayas, mangos, cocoa and coffee beans.

Rainforest's also provide us with many medicinal products. According to the U.S. National Cancer Institute, 70% of plants useful in the treatment of cancer are found only in rainforests. Rainforest plants are also used in the creation of muscle relaxants, steroids, and insecticides. They are used to treat asthma, arthritis, malaria, heart disease, and pneumonia.

The importance of rainforest species in public health is even more incredible considering that less than one percent of rainforest species have been analyzed for their medicinal value.

Threats of Rainforests:

Rainforest's are disappearing at an alarmingly fast pace, largely due to human development over the past few centuries. Once covering 14% of land on Earth, rainforest's now make up only 6%. Since 1947, the total area of tropical rainforests has probably been reduced by more than half. Many biologists expect rainforests will lose 5-10% of their species each decade. Rampant deforestation could cause many important rainforest habitats to disappear completely within the next hundred years.

Such rapid habitat loss is due to the fact that 100 acres of rainforest are cleared every minute for agricultural and industrial development. In the Pacific Northwest's rainforest's, logging companies cut down trees for timber while paper industries use the wood for pulp. In the Amazon rainforest, large-scale agricultural industries, such as cattle ranching, clear huge tracts of forests for arable land. In the Congo rainforest, roads and other infrastructure development have reduced habitat and cut off migration corridors for many rainforest species. Throughout both the Amazon and Congo, mining and logging operations clear-cut to build roads and dig mines. Some rainforests are threatened by massive hydroelectric power projects, where dams flood acres of land. Development is encroaching on rainforest habitats from all sides.

Source:

National Geographic Resource Library

Dedicated to George Edward Hilbert

www.ingramcontent.com/pod-product-compliance
Lightning Source LLC
Chambersburg PA
CBHW050848010526

44107CB00017BA/1212